Engineering Your Way to NASA: A Guide for Aspiring Space Professionals

By Silas Meadowlark

Index

- Exploring the Cosmos
  - The History of Space Exploration
  - The Role of Engineers in Space Missions
  - The Importance of STEM Education

- Fundamental Engineering Principles
  - Mechanics and Dynamics
  - Thermodynamics and Fluid Dynamics
  - Electrical and Electronics Engineering

- Aerospace Engineering Specializations
  - Aerodynamics and Propulsion
  - Structural Design and Materials Science
  - Avionics and Control Systems

- The NASA Application Process
  - Understanding the Hiring Requirements
  - Preparing a Competitive Resume
  - Acing the Interview

- Internships and Co-op Opportunities
  - Identifying Relevant Internship Programs
  - Securing Internships at NASA and Partner Organizations
  - Maximizing the Learning Experience

- Networking and Mentorship
  - Joining Professional Organizations
  - Building Relationships with Industry Experts
  - Seeking Mentorship from Experienced Professionals

- Developing Leadership and Communication Skills
  - Effective Project Management Techniques

- Public Speaking and Presentation Skills
- Teamwork and Collaboration

- Advanced Degree Programs and Research Opportunities
  - Master's and Doctoral Degree Options
  - Participating in NASA-Funded Research Projects
  - Securing Funding and Grants

- The Role of Robotics and Automation in Space Exploration
  - Unmanned Aerial Vehicles (UAVs) and Drones
  - Robotic Manipulators and Rovers
  - Autonomous Systems and Artificial Intelligence

- Sustainability and Environmental Considerations in Space Missions
  - Renewable Energy Sources for Space Habitats
  - Resource Management and Recycling in Space
  - Mitigating the Environmental Impact of Space Exploration

- International Collaboration and the Global Space Industry
  - Multinational Space Programs and Partnerships
  - Navigating Cultural Differences and Language Barriers
  - Exploring Opportunities Abroad

- The Future of Space Exploration
  - Emerging Technologies and Innovations
  - Manned Missions to Mars and Beyond
  - Commercialization and Privatization of Space

- Overcoming Challenges and Setbacks
  - Dealing with Failures and Mistakes
  - Resilience and Perseverance in the Face of Adversity
  - Adapting to Changes and Uncertainties

- Work-Life Balance and Personal Well-Being
  - Maintaining Physical and Mental Health
  - Achieving a Healthy Work-Life Integration

- Fostering a Supportive Professional Network
- Inspiring the Next Generation of Space Professionals
  - Mentoring and Outreach Programs
  - Encouraging STEM Education in Schools
  - Promoting Diversity and Inclusion in the Space Industry

Exploring the Cosmos

The History of Space Exploration

Ah, the grand odyssey of space exploration - a narrative woven with triumph and tragedy, dreams and daring. From the ancient musings of stargazers to the modern day marvels of interstellar voyages, the story of humanity's conquest of the cosmos is a proof to our boundless curiosity and insatiable thirst for knowledge. Let's begin on this celestial journey, where each step has carved a path toward the stars.

The dawn of space exploration can be traced back to the visionary thinkers of the past, whose imaginations soared far beyond the earthly confines. Pioneers like Konstantin Tsiolkovsky and Robert Goddard laid the groundwork for the development of rockets and spacecraft, paving the way for the monumental achievements that would soon follow. The mid 20th century witnessed the iconic space race between the United States and the Soviet Union, a titanic clash of superpowers that captivated the world and pushed the boundaries of human ingenuity.

The crowning glory of this era was the Apollo program, where Neil Armstrong's historic first step on the lunar surface ignited a global sensation and marked a central moment in human history. But the exploration of the cosmos did not end there. The launch of the Hubble Space Telescope, the robotic exploration of the solar system, and the establishment of the International Space Station have all contributed to our ever expanding understanding of the universe and our place within it.

Today, the future of space exploration holds even greater promise, with private companies and international collaborations pushing the boundaries of what's possible. From the development of reusable launch systems to the tantalizing prospect of human missions to Mars, the horizon of space exploration continues to expand, beckoning us to figure out the mysteries that lie beyond our earthly abode.

The Role of Engineers in Space Missions

In the grand blend of space exploration, the unsung heroes are the engineers who breathe life into the dreams of space enthusiasts. These visionary problem solvers, armed with their technical prowess and inventive spirit, are the architects of humanity's journey to the stars.

From the earliest days of rocketry to the cutting edge advancements of modern day spacecraft, engineers have played a central role in shaping the course of space exploration. Their expertise in fields such as aerodynamics, structural design, propulsion systems, and computer science has been instrumental in overcoming the daunting challenges that come with venturing into the unknown.

Take the design and construction of space vehicles, for instance. Engineers must meticulously consider the constraints of weight, materials, and aerodynamics to create vessels that can withstand the rigors of launch, traverse the vastness of space, and safely return to Earth. Their mastery of mechanics, thermodynamics, and electrical engineering is essential in ensuring the reliability and performance of these technological marvels.

But the role of engineers extends far beyond the confines of the spacecraft. They are vital in the development of cutting edge instruments and sensors that collect indispensable data from space, uncovering new frontiers of scientific discovery. Their contributions to the design and operation of robotic probes, orbiting telescopes, and planetary rovers have revolutionized our understanding of the cosmos.

Moreover, engineers are the driving force behind the development of the complex infrastructure that supports space missions, from ground control systems to communication networks. Their ability to tackle complex problems, think creatively, and collaborate effectively is the backbone of the space industry's success.

As we venture deeper into the cosmos, the demand for skilled engineers will only continue to grow. These unsung heroes will be the architects of our future in space, shaping the way we explore, discover, and ultimately, expand our horizons beyond the Earth.

The Importance of STEM Education

In the ever evolving situation of space exploration, the foundation for success lies in the cultivation of a vigorous and diverse talent pool. At the heart of this endeavor is the critical importance of STEM (Science, Technology, Engineering, and Mathematics) education, a clarion call that connects across the global space industry.

As the complication of space missions continues to escalate, the demand for skilled professionals versed in the scientific and technical disciplines has never been more pressing.

From designing cutting edge spacecraft to developing the algorithms that guide robotic probes, the space industry requires a steady influx of inventive thinkers, problem solvers, and visionaries.

STEM education plays a crucial role in nurturing these essential skills, instilling in students a deep understanding of the fundamental principles that underpin the exploration of the cosmos. Whether it's mastering the intricacies of aerodynamics, delving into the mysteries of astrophysics, or grappling with the challenges of software engineering, a strong STEM foundation equips aspiring space professionals with the tools they need to thrive in this dynamic and ever evolving field.

But the significance of STEM education extends far beyond the immediate needs of the space industry. By inspiring a new generation of scientific thinkers and tech savvy innovators, we are cultivating a workforce that will shape the future of humanity's exploration and understanding of the universe. These individuals will be the trailblazers who push the boundaries of what's possible, revealing new frontiers of discovery and shaping the course of space exploration for decades to come.

Moreover, the importance of STEM education extends beyond the confines of the space industry, with its principles and practices serving as a powerful catalyst for innovation and problem solving across a wide range of disciplines. By promoting a culture of curiosity, critical thinking, and collaborative problem solving, STEM education enables individuals to tackle the complex challenges that face our world, whether it's addressing climate change, developing sustainable energy solutions, or improving healthcare outcomes.

As we stand on the precipice of a new era of space

exploration, the vital role of STEM education in nurturing the next generation of space professionals cannot be overstated. By investing in the cultivation of these essential skills, we are not only securing the future of the space industry but also equipping a new generation of innovators and problem solvers to shape the destiny of humanity itself.

Fundamental Engineering Principles

Mechanics and Dynamics

Buckle up, folks, because we're about to dive headfirst into the wild, wild world of mechanics and dynamics! This is the foundation upon which the entire universe of engineering rests, so you better believe we're going to give it the attention it deserves. Strap on your thinking caps and get ready to have your minds blown.

First up, let's talk about Newton's Laws of Motion. These bad boys are the building blocks of everything from designing the perfect paper airplane to sending a rocket to the moon. Mastering these principles will make you feel like a superhero, able to predict and control the movements of objects with ease. Trust me, once you really get a handle on Newton, you'll be able to impress your friends, family, and that cute engineer you've been trying to catch the eye of at the local STEM mixer.

Speaking of motion, let's not forget about kinematics and dynamics. These concepts will help you understand the complicated dance of position, velocity, acceleration, and the forces that drive them. Whether you're designing a state of-the art rover for a Mars mission or trying to improve the trajectory of a soccer ball, these principles will be your secret weapons.

And let's not forget about good old statics and structural analysis. These are the unsung heroes of the engineering world, keeping everything from bridges to spacecraft from collapsing under their own weight. Trust me, when you can calculate the exact moment a structure will fail, you'll feel like a total badass.

So buckle up, put your pocket protectors on, and get ready to dive deep into the world of mechanics and dynamics. It may sound daunting, but with a little elbow grease and a whole lot of passion, you'll be soaring through the cosmos in no time.

Thermodynamics and Fluid Dynamics

Ah, thermodynamics and fluid dynamics – the two scientific disciplines that will have you questioning the very nature of the universe. But don't let that intimidate you, my space faring friends. These principles are the key to making accessible the secrets of the cosmos, from the inner workings of rocket engines to the complex atmospheric conditions that govern the weather on distant planets.

First, let's talk about thermodynamics. This is the study of heat, energy, and the mind bending laws that govern their behavior. Think of it as the cosmic dance between temperature, pressure, and the flow of heat – a delicate ballet that engineers must master to design everything from power plants to life support systems for space habitats.

And then there's fluid dynamics, the study of how liquids and gases move and interact. This is the realm of fluid mechanics, where the principles of continuity, momentum, and viscosity reign supreme. Whether you're designing the optimal shape for a spacecraft's aerodynamic hull or calculating the flow of blood through the human body, fluid dynamics will be your indispensable ally.

But don't think these subjects are all about dry equations and complex formulas. Oh, no, my friends. Thermodynamics and fluid dynamics are where the rubber meets the road, where theory and practice come together in a spectacular display of engineering prowess. Imagine the thrill of watching a rocket launch, knowing that the precise calculations you made in the lab were the key to its successful liftoff. Or the satisfaction of seeing your design for a life support system keep astronauts alive and thriving in the harsh environment of space.

So, strap on your thinking caps, grab your slide rules, and get ready to dive into the mesmerizing world of thermodynamics and fluid dynamics. Trust me, once you've mastered these principles, the sky - or rather, the cosmos - will be the limit!

Electrical and Electronics Engineering

Attention, all you aspiring space cadets! Are you ready to use the power of electricity and electronics to conquer the final frontier? Because let me tell you, this is where the real magic happens - the stuff that turns science fiction into reality.

First things first, let's talk about the fundamentals of electrical engineering. This is where you'll learn the secrets of circuits, currents, and conductors - the building blocks of everything from spacecraft avionics to the handheld gadgets that keep astronauts connected to mission control. Get ready to dive deep into the world of Ohm's Law, resistance, and capacitance – the stuff that makes the universe tick.

But it doesn't stop there, my friends. Electronics engineering is where the real fireworks happen. This is where you'll learn to use the power of semiconductors, integrated circuits, and microprocessors – the heart and soul of the digital revolution. Imagine designing the detailed control systems that keep a Mars rover rolling across the dusty surface of the red planet, or the communication networks that allow mission control to stay in touch with our intrepid explorers in the depths of space.

And let's not forget about power systems, the lifeblood that keeps everything running smoothly. From solar panels and fuel cells to batteries and power converters, these are the unsung heroes that make space exploration possible. Mastering the principles of power generation, storage, and distribution will have you feeling like a cosmic electrician, keeping the engines of discovery firing on all cylinders.

So, what are you waiting for? Strap on your goggles, grab your soldering iron, and get ready to dive into the electrifying world of electrical and electronics engineering. With these skills under your belt, the sky – or rather, the entire universe – will be your playground.

Aerospace Engineering Specializations

Aerodynamics and Propulsion

In the realm of aerospace engineering, the field of aerodynamics and propulsion stands as a cornerstone of space exploration. This captivating discipline explores into the complex dance of air and spacecraft, revealing the secrets of lift, drag, and the high powered systems that propel us beyond the confines of Earth's atmosphere.

Aspiring space professionals must embrace the art of fluid dynamics, understanding the complex interplay of pressure, velocity, and temperature that dictates the behavior of gases and liquids. From the design of sleek, aerodynamic fuselages to the engineering of cutting edge propulsion systems, this specialization demands a deep understanding of the principles governing the motion of fluids.

The development of advanced propulsion technologies, such as chemical rockets, ion engines, and nuclear powered systems, is a persistent pursuit that pushes the boundaries of human ingenuity. Mastering the intricacies of combustion, plasma physics, and exotic fuel sources is the key to uncovering the next generation of space travel. Budding aerospace engineers must be prepared to tackle the challenges of thrust, efficiency, and environmental impact head on, paving the way for ever bolder missions to the farthest reaches of the cosmos.

Structural Design and Materials Science

The rugged and dependable structures that carry our dreams into the stars are the province of structural design and materials science. This captivating field encompasses the engineering of airframes, spacecraft, and the critical components that must withstand the immense stresses of launch, atmospheric entry, and the harsh environments of space.

Aspiring space professionals must be well versed in the properties and behavior of advanced materials, from the lightweight yet sturdy composites that form the backbone of modern aircraft to the high temperature alloys and ceramics essential for the protection of spacecraft during fiery atmospheric re entry. The ability to analyze structural loads, anticipate fatigue and failure modes, and design strong, lightweight systems is of great importance in this specialization.

Beyond the physical structure, materials science also plays a essential role in the development of cutting edge technologies for space exploration. From the thermal protection systems that shield astronauts and equipment from the blistering heat of launch to the radiation resistant components that safeguard sensitive electronics, the innovations in this field are the foundation upon which the future of space travel is built. Aspiring aerospace engineers must embrace the challenge of constantly pushing the boundaries of what's possible, crafting the structures and materials that will carry humanity to the stars.

Avionics and Control Systems

In the realm of aerospace engineering, the detailed dance of avionics and control systems is the heartbeat that guides our journey through the cosmos. This captivating specialization encompasses the design, integration, and management of the sophisticated electronic systems that allow spacecraft and aircraft to navigate, communicate, and autonomously maintain their course, even in the face of the most demanding conditions.

Aspiring space professionals must be adept at the principles of electrical and electronics engineering, mastering the intricacies of sensors, processors, and communication networks. From the precise guidance systems that steer spacecraft to the redundant power management systems that ensure mission critical operations, every component in this field plays a vital role in the success of a space exploration venture.

The field of avionics and control systems is constantly evolving, driven by the unceasing pursuit of increased reliability, efficiency, and adaptability. Emerging technologies, such as artificial intelligence and machine learning, are transforming the way spacecraft and aircraft are operated, enabling autonomous decision making and adaptive control in the face of unpredictable challenges. Budding aerospace engineers must be prepared to embrace this technological revolution, using the power of advanced electronics and software to push the boundaries of what's possible in space exploration.

The NASA Application Process

Understanding the Hiring Requirements

Navigating the detailed world of NASA's hiring process can be a daunting task, but fear not, aspiring space professionals! The key to success lies in understanding the rigorous requirements and tailoring your application to showcase your exceptional qualifications. NASA seeks individuals who possess a unique blend of technical expertise, problem solving skills, and a genuine passion for space exploration.

First and foremost, a strong educational foundation is a must. NASA typically requires a bachelor's degree in engineering, physics, or a related STEM field, with a stellar academic record to boot. But don't just stop there – many positions also necessitate advanced degrees, such as a master's or doctorate, to truly stand out in the competitive applicant pool.

Technical proficiency is the backbone of any successful NASA engineer. You'll need to demonstrate mastery of core engineering principles, from fluid dynamics and thermodynamics to electrical and control systems. Hands on experience, whether through internships, research projects, or personal tinkering, can provide a critical edge.

But it's not just about the technical know how – NASA values well rounded individuals who can thrive in a collaborative, high pressure environment. Strong communication skills, the ability to work effectively in diverse teams, and a proven track record of problem solving and decision making are all essential qualities.

Lastly, a deep passion for space exploration and a genuine commitment to the mission of NASA are non negotiable. Your enthusiasm and dedication should shine through in every aspect of your application, from your cover letter to your interviews.

Preparing a Competitive Resume

Now that you understand the key requirements, it's time to craft a resume that will make NASA sit up and take notice. Remember, your resume is your first and most important opportunity to showcase your qualifications and stand out from the crowd.

Begin by highlighting your educational achievements, focusing not just on your degrees but also any relevant coursework, research projects, or academic honors that demonstrate your technical prowess. Don't be afraid to showcase your hands on experience, whether it's through internships, co op programs, or personal projects.

Next, dive into your professional experience, emphasizing your contributions and accomplishments in previous roles. Quantify your impact wherever possible, showcasing how your work directly improved processes, saved resources, or drove inventive solutions. Tailor your language to showcase the specific skills and competencies NASA values, such as problem solving, teamwork, and effective communication.

Finally, don't forget to sprinkle in your passion for space exploration and your commitment to the NASA mission. Share any volunteer work, memberships in relevant organizations, or personal achievements that demonstrate your dedication to the field.

Remember, your resume is not just a list of your experiences – it's a chance to tell a compelling story about your journey and your potential to contribute to the incredible work of NASA.

Acing the Interview

You've nailed the resume, and now it's time to shine in the interview. This is your opportunity to showcase your technical expertise, problem solving abilities, and passion for the role, all while navigating the high stakes environment of a NASA interview.

Prepare thoroughly by researching the specific position, the agency's ongoing projects, and the latest advancements in the field. Anticipate the types of questions you might be asked, from technical challenges to behavioral scenarios, and practice your responses until they feel natural and confident.

During the interview, be ready to examine into the nitty gritty of your previous experiences, highlighting the skills and knowledge you've gained. But don't just recite your resume – weave in personal anecdotes and examples that demonstrate your ability to think on your feet and tackle complex problems.

Most importantly, let your enthusiasm for space exploration shine through. Share your dreams, your fascination with the cosmos, and your unwavering commitment to the NASA mission. Interviewers want to see that you're not just a skilled engineer, but a true visionary who can contribute to the agency's new work.

Remember, the interview is not just a one way street – be prepared to ask insightful questions that showcase your understanding of the role and your genuine interest in the organization. This will not only impress your interviewers but also help you determine if the position is the right fit for your aspirations.

Embrace the challenge, stay poised under pressure, and let your passion for space exploration shine through. With the right preparation and mindset, you'll be well on your way to securing your dream job at NASA.

Internships and Co op Opportunities

Identifying Relevant Internship Programs

Ah, the sweet smell of opportunity! As an aspiring space professional, securing the right internship can be the launchpad to your stratospheric career. But with so many options out there, how do you find the perfect fit? Well, buckle up, because we're about to take a comprehensive study into the world of out of-this world internships.

First and foremost, let's talk about NASA. The agency's internship programs are the holy grail for space enthusiasts, but don't be intimidated. These opportunities are highly competitive, so you'll need to bring your A game. Start by scouring the NASA website for listings – from engineering to astrophysics, they've got a little something for everyone. But don't stop there! Explore internships with private aerospace companies, university research labs, and even international space agencies. The more options you explore, the better your chances of finding the perfect mission control for your talents.

Now, here's a pro tip: don't just focus on the big names. Some of the most exciting and novel work is happening at smaller, lesser known organizations. So, be sure to cast a wide net and don't be afraid to think outside the launch pad. Who knows, you might just stumble upon the next SpaceX or Blue Origin in the making.

Remember, when it comes to internships, it's not just about the work – it's about the connections you make. Reach out to alumni, professors, and industry professionals to get the inside scoop on the best opportunities. Networking is the cosmic glue that holds the space industry together, so start building those relationships now.

Securing Internships at NASA and Partner Organizations

Now that you've got a handle on the internship setting, it's time to put your game face on and start the application process. But before you start firing off those resumes, let's talk strategy.

First and foremost, do your homework. Research the specific requirements and qualifications for each internship program you're considering. Some may prioritize technical skills, while others might be looking for leadership and communication chops. Tailor your application materials to showcase your strengths and how they match with the program's needs.

Speaking of application materials, let's talk about your resume. This is your chance to blast off and leave a lasting impression. Highlight your STEM accomplishments, relevant coursework, and any hands on experience you've gained. But don't just list your achievements – tell a story. Weave in your passion for space exploration and your unique perspective. After all, you're not just another candidate – you're a cosmic pioneer in the making.

And don't forget about the interview process. These opportunities are highly competitive, so you'll need to bring your A game. Practice your responses, polish your public speaking skills, and be prepared to showcase your problem solving abilities. Remember, the interviewers want to see not just your technical prowess, but your ability to think on your feet and thrive in a dynamic, fast paced environment.

Finally, don't be afraid to get creative. Sometimes, the traditional application process just doesn't cut it. Maybe you could reach out to current interns and ask for their understanding, or even put together a personalized video pitch highlighting your skills and enthusiasm. Think outside the launch pad, and you might just find yourself blasting off to the space of your dreams.

Maximizing the Learning Experience

Congratulations, you've landed that out of-this world internship! But the real work has just begun. Now it's time to make the most of your cosmic opportunity.

First and foremost, be a sponge. Soak up every bit of knowledge and experience you can. Attend every workshop, shadow every engineer, and ask a million questions. Don't be afraid to dive headfirst into projects that might be outside your comfort zone – after all, that's where the real learning happens.

And speaking of projects, make sure to get your hands dirty. Volunteer for tasks that allow you to apply your technical skills and problem solving abilities. Whether it's designing a new spacecraft component or developing an original software algorithm, the more you can contribute, the more you'll get out of the experience.

But it's not just about the work – it's about the connections you make. Network like your career depends on it (spoiler alert: it does). Introduce yourself to industry leaders, pick the brains of seasoned professionals, and forge relationships that could lead to future opportunities. After all, in the space industry, it's not just what you know, but who you know.

Finally, don't forget to have fun! Sure, interning at NASA or a top aerospace company might be intense, but that doesn't mean you can't enjoy the ride. Embrace the camaraderie of your fellow interns, explore the local space themed attractions, and soak in the sheer wonder of being part of something so much bigger than yourself. After all, you're not just an intern – you're a cosmic pioneer in the making.

Networking and Mentorship

Joining Professional Organizations

When it comes to breaking into the space industry, it's not just what you know, but who you know. And the best way to build those all important connections? Get your foot in the door by joining professional organizations. These groups are like a secret handshake for space nerds – they'll open up a world of networking opportunities, industry perceptions, and mentorship that you simply can't get anywhere else.

Start by researching organizations like the American Institute of Aeronautics and Astronautics (AIAA) or the Society of Women Engineers (SWE). These groups offer a abundance of resources, from technical workshops to career development programs. But the real gold is in the connections you'll make. Attend their conferences, volunteer for committees, and don't be afraid to strike up conversations with seasoned pros. You never know when that chance encounter could lead to your big break.

And don't just limit yourself to aerospace specific groups. Broaden your horizons by exploring organizations related to your engineering specialization, like the American Society of Mechanical Engineers (ASME) or the Institute of Electrical and Electronics Engineers (IEEE). The more diverse your network, the better equipped you'll be to tackle the many-sided challenges of space exploration.

Building Relationships with Industry Experts

Networking isn't just about collecting business cards and LinkedIn connections - it's about forging genuine, meaningful relationships with the movers and shakers of the space industry. These industry experts, from veteran astronauts to pioneering engineers, are a rich source of knowledge and experience. And they're often more than willing to share it, if you know how to approach them the right way.

Start by identifying the people you admire most - the ones whose work has inspired you or whose career path you'd love to emulate. Reach out to them directly, whether it's through a professional organization or a personal connection. But don't just ask for a job or an internship - instead, express genuine interest in their work and expertise. Ask thoughtful questions, offer to help with a project, or simply request a few minutes of their time to pick their brain.

Remember, these industry leaders are busy people, so be respectful of their time and schedules. But if you approach them with genuine curiosity and a willingness to learn, you'd be surprised how many of them are eager to pay it forward and mentor the next generation of space professionals. And who knows - those connections could be the key to revealing the door to your dream job at NASA.

Seeking Mentorship from Experienced Professionals

In the high stakes, high pressure world of space exploration, having a trusted mentor in your corner can make all the difference. These experienced professionals, whether they're seasoned engineers or veteran astronauts, can offer indispensable guidance, support, and a wealth of knowledge that can't be found in any textbook.

But finding the right mentor isn't always easy. It's not just about finding someone with impressive credentials - it's about finding someone with whom you share a genuine connection and mutual respect. Start by tapping into your existing network, whether it's through professional organizations, internships, or even your university's alumni network. Reach out to people whose work or career path has inspired you, and express your interest in learning from their experiences.

Once you've identified a potential mentor, be proactive in nurturing the relationship. Set up regular check ins, whether it's a monthly coffee or a quick Zoom call. Come prepared with thoughtful questions and a genuine willingness to listen and learn. And don't be afraid to ask for specific advice or feedback - a good mentor will appreciate your initiative and dedication.

Remember, the mentorship relationship is a two way street. Be sure to offer your own unique perspectives and skills, and find ways to support your mentor's work or goals. The more you can establish a mutually beneficial partnership, the stronger the bond will become - and the more it will pay dividends in your own career development.

Developing Leadership and Communication Skills

Effective Project Management Techniques

In the high stakes world of space exploration, successful project management is the unsung hero that makes the impossible possible. As an aspiring space professional, honing your project management skills is essential to navigating the complex web of deadlines, budgets, and team dynamics that define NASA's mission driven environment. It's not just about creating Gantt charts and holding status meetings - it's about cultivating the mindset of a true space leader.

Start by embracing the art of task prioritization. Forget the traditional to do lists - think of your project as a tightrope, and each task as a precarious step. Identify the critical path, the make or-break milestones that will determine the success or failure of your mission. Ruthlessly eliminate anything that doesn't directly support those key objectives, even if it means saying no to the boss's pet idea about adding a holographic unicorn to the spacecraft's control panel.

Next, promote a culture of transparent communication. In the high pressure cauldron of space engineering, egos and silos can quickly derail even the most well planned project.

Break down those barriers by creating an environment where team members feel allowed to speak up, share concerns, and collaborate freely. Hold regular "lessons learned" sessions, where you celebrate successes and dissect failures with equal enthusiasm. After all, the path to the stars is paved with the lessons of those who came before.

Finally, master the art of risk management. In the ever changing situation of space exploration, expecting the unexpected is the only way to stay afloat. Develop a keen eye for potential roadblocks, and always have a contingency plan ready to deploy at a moment's notice. When the inevitable curveball comes hurtling your way, be the calm, collected leader who guides your team through the storm, emerging stronger and more resilient than ever before.

Public Speaking and Presentation Skills

In the high stakes world of space exploration, the ability to captivate an audience is just as critical as your technical prowess. Whether you're pitching a revolutionary new spacecraft design to NASA executives or delivering a TED talk on the future of robotic exploration, mastering the art of public speaking and presentation skills can make the difference between soaring success and crashing and burning.

Start by embracing the power of storytelling. Gone are the days of dry, data heavy presentations that put your audience to sleep faster than a trip to the International Space Station. Instead, craft a narrative that draws your listeners in, transporting them to the heart of your mission. Use vivid imagery, personal anecdotes, and a dash of humor to make your message soar.

Next, employ the power of your body language. Your posture, gestures, and eye contact can make or break your performance. Stand tall, make direct eye contact with your audience, and use your hands to underline your key points. Avoid the temptation to hide behind a podium or fidget nervously - own the stage, and your audience will be captivated.

Finally, conquer your fears and embrace the power of practice. Public speaking can be a daunting prospect, even for the most seasoned space professionals. But the more you put yourself out there, the more comfortable you'll become. Seek out every opportunity to speak, from team meetings to industry conferences. Record yourself, get feedback, and continuously refine your approach until you're delivering presentations that leave your audience in awe.

Teamwork and Collaboration

In the high stakes world of space exploration, no one achieves greatness alone. From designing the next generation of rocket engines to developing cutting edge scientific instrumentation, the most inventive feats in the cosmos are the product of fluid teamwork and collaboration.

Start by embracing the power of diversity. Surround yourself with individuals who bring a wide range of expertise, perspectives, and problem solving approaches to the table. Resist the temptation to surround yourself with "yes men" who simply echo your own ideas - instead, seek out those who will challenge your assumptions and push you to think beyond the boundaries of the status quo.

Next, grow a culture of mutual respect and trust. In the high pressure cauldron of space engineering, tensions can quickly flare, and egos can clash. Break down those barriers by promoting an environment where everyone feels heard, valued, and equipped to contribute. Hold regular team building exercises, celebrate small wins, and never underestimate the power of a well timed coffee break to recharge and reconnect.

Finally, hone your conflict resolution skills. In the world of space exploration, where lives and multi million-dollar budgets hang in the balance, the ability to navigate interpersonal conflicts with grace and diplomacy is of great importance. Learn to listen actively, seek to understand different viewpoints, and prioritize the greater good over individual agendas. When the inevitable disagreements arise, be the voice of reason that guides your team through the storm, emerging stronger and more united than ever before.

Advanced Degree Programs and Research Opportunities

Master's and Doctoral Degree Options

If you've got your sights set on the stars, don't just settle for a bachelor's degree – reach for the cosmos with an advanced degree program! Sure, you could coast along with a standard engineering diploma, but where's the fun in that? Be the one who shows up to the office in a spacesuit, demanding a raise because you've got a PhD in Quantum Astrophysics. Trust me, it'll get you places.

Now, I know what you're thinking – "But wait, don't I have to sell my soul and become a hermit for 10 years to get an advanced degree?" Not necessarily, my friend. With the right strategy, you can earn that shiny new diploma while still maintaining a social life (or at least, pretending to have one).

Start by scoping out the top notch aerospace engineering programs at universities across the country. Look for schools with strong ties to NASA and other space agencies – those are the real goldmines. Bonus points if the campus has a secret underground lab where they're working on warp drive technology. Just make sure to keep that under wraps, or they might revoke your student ID.

Once you've found your dream program, get ready to dive

headfirst into a world of research, endless problem sets, and more caffeine than a space shuttle launch. But hey, at least you'll be in good company – your fellow students will be just as sleep deprived and slightly unhinged as you are. Embrace the madness, and you might just come out the other side with a degree that'll make your parents proud (or at least slightly less confused about your career choices).

Participating in NASA Funded Research Projects

If you really want to take your space career to the next level, there's no better way than getting involved in NASA funded research projects. Imagine it – you, standing in front of a room full of rocket scientists, presenting your pioneering findings on the optimal angle of attack for a Mars rover's solar panels. It's a dream come true, and it could be yours if you play your cards right.

The key is to start networking, networking, and more networking. Get involved with professional organizations, attend conferences, and make friends with the kind of people who can get your foot in the door at NASA. Sure, it might feel a little like you're trying to join the secret space cadet club, but trust me, it's worth it.

Once you've made those all important connections, start looking for opportunities to join research teams working on NASA funded projects. These can range from studying the effects of microgravity on plant growth to developing new materials for spacecraft construction. The possibilities are endless, and the experience you'll gain is truly out of this world (pun very much intended).

And let's not forget about the pure, unadulterated joy of working with the best and brightest minds in the space industry. Imagine bouncing ideas off astronauts, engineers, and scientists who have literally reached for the stars. It's the kind of thing that'll make you feel like you're living in a sci fi movie, except with way fewer alien invasions (probably).

Securing Funding and Grants

Alright, let's talk about the elephant in the room – money. We all know that pursuing an advanced degree or getting involved in NASA funded research isn't exactly cheap. But fear not, my space faring friends, because there are plenty of ways to score the funding you need to make your dreams a reality.

First and foremost, start exploring the world of grants and scholarships. Trust me, there's a veritable galaxy of funding opportunities out there, just waiting to be discovered. Do your research, get creative with your applications, and don't be afraid to think outside the box. After all, who knows – your essay about how you plan to use your degree to colonize Pluto might just be the winning ticket.

And let's not forget about good old fashioned networking. Remember all those connections you made earlier? Well, now's the time to put them to work. Reach out to industry experts, alumni associations, and even your university's financial aid office – they might just have the inside scoop on funding sources you never even knew existed.

Of course, if all else fails, there's always the option of becoming a professional cat video influencer and using your newfound fame to crowdfund your way to a PhD. Hey, stranger things have happened, right? Just don't forget to include a stretch goal for a zero gravity cat hammock in your campaign.

The Role of Robotics and Automation in Space Exploration

Unmanned Aerial Vehicles (UAVs) and Drones

As humanity's quest for interstellar domination continues, the role of unmanned aerial vehicles (UAVs) and drones has become increasingly critical in the grand scheme of space exploration. These high tech aerial marvels, once the stuff of science fiction, have now become indispensable tools in our arsenal of intergalactic conquest.

Gone are the days when space missions were solely the domain of human astronauts. Today, these autonomous flying machines have become the unsung heroes, serving as our eyes and ears in the vast, unforgiving expanse of the cosmos. Imagine a future where swarms of nimble, AI powered drones dart between celestial bodies, gathering data, conducting reconnaissance, and paving the way for manned missions. It's a scene straight out of a futuristic blockbuster, but it's quickly becoming our reality.

These versatile aerial platforms, equipped with cutting edge sensors and advanced navigation systems, can navigate the treacherous environments of alien worlds, where the mere thought of a human presence would be akin to a lemming leaping off a cliff. From mapping the rugged terrain of distant planets to monitoring the health of spacecraft, UAVs and drones have become indispensable tools in the toolbox of the modern space explorer.

As we push the boundaries of space exploration, these robotic marvels will continue to evolve, becoming more capable, more autonomous, and more integrated into the fabric of our interstellar endeavors. Who knows, perhaps one day we'll see entire fleets of drones, working in tandem with human crews, as we venture forth to conquer the unknown.

Robotic Manipulators and Rovers

Ah, the humble robot – once the stuff of 1950s science fiction, now the unsung heroes of modern space exploration. These tireless, constant machines have become the silent sentinels, venturing where no human dare tread, performing tasks that would make even the bravest astronaut quiver in their spacesuit.

Take, for instance, the robotic manipulators – those dexterous, mechanical arms that can delicately handle the most fragile of instruments, or the mighty rovers that traverse the unforgiving terrain of distant planets, gathering data and samples as they go. These robotic wonders have become the indispensable extensions of our own capabilities, allowing us to explore the cosmos in ways that were once deemed impossible.

But these are no mere mindless automatons – no, these are the result of cutting edge engineering, the culmination of decades of research and innovation. Each joint, each sensor, each line of code is carefully crafted to push the boundaries of what is possible, to overcome the challenges that nature has thrown at us in our quest for interstellar domination.

And as we continue to push the limits of what these robotic marvels can do, the future of space exploration becomes ever more tantalizing. Imagine a day when entire fleets of autonomous rovers, guided by the latest in artificial intelligence, scour the surface of Mars, uncovering the secrets of the red planet. Or picture a swarm of robotic manipulators, working in perfect harmony, assembling the next generation of space stations in the void of the cosmos.

Yes, the robots have come a long way, and their role in the future of space exploration is only going to grow more vital, more vital, and more awe inspiring. So, the next time you see a rover trundling across the surface of a distant world, or a robotic arm delicately handling a priceless sample, take a moment to marvel at the ingenuity of the human mind and the endless possibilities of the machine.

Autonomous Systems and Artificial Intelligence

In the grand symphony of space exploration, the rising crescendo of autonomous systems and artificial intelligence has become a veritable symphony of innovation. Gone are the days when human hands were the sole orchestrators of our interstellar endeavors – now, the machines have taken center stage, and they're conducting a performance that would make even the most seasoned maestro green with envy.

Imagine a future where entire space missions are carried out without a single human astronaut setting foot outside of Earth's atmosphere. Where autonomous systems, guided by the lightning fast decision making of artificial intelligence, navigate the treacherous void of the cosmos, conducting experiments, gathering data, and even repairing and maintaining the very spacecraft that carries them.

It's a future that is rapidly becoming a reality, thanks to the tireless efforts of engineers, researchers, and visionaries who are pushing the boundaries of what's possible. From self piloting spacecraft to robotic repair crews, the integration of autonomous systems and artificial intelligence into the fabric of space exploration is nothing short of a revolution.

But this is no mere technological parlor trick – the implications of these advancements are intense, and the potential benefits are staggering. Imagine a world where the risks of manned space missions are drastically reduced, where the exploration of the solar system and beyond is no longer constrained by the fragile limitations of the human body.

And as these autonomous systems and AI powered technologies continue to evolve, the future of space exploration becomes ever more exciting. Perhaps one day, we'll see entire colonies on distant planets, managed and maintained by a symphony of robotic caretakers, guided by the wisdom of artificial intelligence. Or maybe, just maybe, we'll witness the birth of a new era of space exploration, where the line between human and machine becomes so blurred that it's impossible to tell where one ends and the other begins.

Whatever the future holds, one thing is certain: the role of autonomous systems and artificial intelligence in the conquest of the cosmos is only going to grow more vital, more essential, and more awe inspiring. So, buckle up, space cadets – the future is here, and it's a wild ride.

Sustainability and Environmental Considerations in Space Missions

Renewable Energy Sources for Space Habitats

As humanity sets its sights on expanding its presence beyond the confines of Earth, the challenge of powering our extraterrestrial outposts has become a pressing concern. Conventional fossil fuels are simply not a viable option for the long term sustainability of space missions. Fortunately, the realm of renewable energy has emerged as a game changer, offering a multitude of inventive solutions to fuel our exploration of the cosmos.

Solar energy, the perennial favorite, has found a natural home in the vacuum of space. High efficiency solar panels, capable of converting the abundant solar radiation into usable electricity, have become the backbone of many space habitats. These advanced photovoltaic systems, coupled with state of-the art energy storage technologies, ensure a reliable and continuous supply of power, even during the darkness of lunar nights or the unpredictable nature of solar activity.

But the ingenuity of space engineers doesn't stop there. Cutting edge wind turbines, designed to apply the unique atmospheric conditions found on other planetary bodies, have also been explored as a means of generating renewable energy. Imagine the sights of towering wind farms on the surface of Mars, channeling the planet's fierce gusts to power our future settlements.

Beyond these well known renewable sources, the exploration of inventive solutions, such as microbial fuel cells and advanced nuclear reactors, has opened up new frontiers in extraterrestrial energy production. These technologies, while still in their infancy, hold the promise of delivering clean, efficient, and adjustable power to support the long term sustainability of our space based endeavors.

Resource Management and Recycling in Space

As we venture deeper into the cosmos, the concept of a closed loop resource management system becomes very important. The harsh realities of operating in the vacuum of space dictate a fundamental shift in our approach to resource utilization and waste management. Embracing the principles of the circular economy, space engineers are leading the charge in revolutionizing the way we view and manage the precious resources at our disposal.

Water, a precious commodity in the vastness of space, has become the focal point of many original recycling initiatives. Advanced water purification systems, capable of reclaiming and reusing every drop, have become essential components of space habitats. The ability to recycle and recirculate water not only reduces the need for frequent resupply missions but also minimizes the environmental impact of our space based activities.

Similarly, the recovery and reuse of materials, from metals to plastics, have become a important aspect of sustainable space exploration. 3D printing technologies, combined with sophisticated recycling processes, enable the conversion of discarded components into new, repurposed parts and equipment. This circular approach to resource management not only reduces waste but also encourages a culture of innovation and self sufficiency among space faring nations and private enterprises.

The ultimate goal of this overall approach to resource management is to create a truly self sustaining community, where the inputs and outputs of space missions are carefully balanced and optimized for maximum efficiency and minimum environmental impact. As we push the boundaries of human presence in space, the lessons learned from these pioneering efforts will undoubtedly shape the future of sustainable development on Earth as well.

Mitigating the Environmental Impact of Space Exploration

The pursuit of space exploration has long been synonymous with technological advancement and scientific discovery. However, as our ambitions grow and our presence in the cosmos expands, it has become increasingly important to consider the environmental implications of our activities beyond Earth's atmosphere.

One of the primary concerns is the impact of rocket launches and spacecraft operations on the delicate balance of our planet's atmosphere. The emission of greenhouse gases, particulate matter, and other pollutants during launch and reentry can contribute to the degradation of the ozone layer and the acceleration of climate change. Space agencies and private companies are now at the forefront of developing cleaner, more efficient propulsion systems that minimize the environmental footprint of their missions.

Furthermore, the management of space debris has emerged as a important issue that demands the attention of the global space community. Defunct satellites, discarded rocket stages, and other man made objects orbiting the Earth pose a significant threat to active spacecraft and the long term sustainability of space operations. Novel solutions, such as advanced debris tracking systems and specialized debris removal technologies, are being explored to mitigate this growing problem.

Looking beyond the immediate impact on our own planet, the potential environmental consequences of human activities on other celestial bodies must also be considered. The principle of planetary protection, which aims to prevent the contamination of extraterrestrial environments, has become a guiding force in the design and execution of space missions. Rigorous sterilization protocols, the development of containment measures, and the implementation of responsible exploration practices are all vital elements in preserving the pristine nature of the environments we encounter in our cosmic voyages.

As we continue to push the boundaries of space exploration, the integration of sustainable practices and environmental consciousness into every aspect of our space based endeavors will be of great importance. By embracing a complete approach to mitigating the environmental impact of our space faring activities, we can ensure that our exploration of the cosmos is not only technologically and scientifically pioneering but also environmentally responsible and forward thinking.

International Collaboration and the Global Space Industry

Multinational Space Programs and Partnerships

In today's interconnected world, the space industry has embraced a global perspective, with nations coming together to push the boundaries of human exploration and scientific discovery. Multinational space programs have become the norm, as countries pool their resources, expertise, and ambitions to tackle the challenges of space travel and research. From the iconic International Space Station, where astronauts from around the world work side by side, to joint missions that apply complementary capabilities, international collaboration has become the lifeblood of the space industry.

At the forefront of this global collaboration is NASA, which has forged partnerships with space agencies from Europe, Russia, Japan, Canada, and beyond. These partnerships have not only facilitated inventive achievements but have also nurtured cultural exchange, technological innovation, and a shared sense of purpose. By working together, space faring nations can share the immense costs, risks, and rewards of pushing the boundaries of what's possible beyond our planet.

But navigating these multinational endeavors is not without its complexities. Differences in language, cultural norms, and bureaucratic processes can present unique challenges that must be overcome through effective communication, mutual understanding, and a willingness to compromise. Successful space professionals must possess the adaptability and diplomatic skills to navigate these complex collaborations, ensuring that the collective vision remains focused and the mission objectives are achieved.

Navigating Cultural Differences and Language Barriers

As the space industry becomes increasingly global, aspiring professionals must be prepared to work with individuals from diverse cultural backgrounds and linguistic traditions. Effective communication and cross cultural competence are essential for success in this international arena.

Understanding and respecting cultural differences is of great importance. What may be considered acceptable behavior or communication style in one country may be seen as rude or inappropriate in another. Successful space professionals must be attuned to these nuances, adapting their communication and behavior to ensure smooth collaboration and mutual understanding.

Language barriers can also pose a significant challenge, particularly in mission critical scenarios where clear and precise communication is essential. Mastering the art of effective cross language dialogue, utilizing translation services, and developing language skills can be indispensable assets. Additionally, cultivating cultural sensitivity and a willingness to learn from one's international colleagues can nurture an environment of trust, respect, and shared success.

Overcoming these cultural and linguistic hurdles requires a combination of technical expertise, emotional intelligence, and a genuine curiosity to understand and appreciate diverse perspectives. By embracing the richness of global collaboration, aspiring space professionals can uncover a world of opportunities and contribute to new achievements that transcend national boundaries.

Exploring Opportunities Abroad

The global nature of the space industry presents a wealth of opportunities for aspiring professionals to explore beyond the borders of their home countries. From internships and cooperative programs to full time employment, the international space terrain offers a diverse array of possibilities for those seeking to expand their horizons and gain valuable experience.

Aspiring space professionals should be proactive in researching and identifying internship and co op programs offered by international space agencies, research institutions, and private aerospace companies. These immersive experiences can provide indispensable exposure to different working cultures, new technologies, and unique project challenges, ultimately improving one's technical skills and broadening their perspective.

For those seeking long term career opportunities, the global space industry presents a wealth of options. With multinational space agencies and private companies establishing a presence around the world, the ability to work abroad can open doors to diverse projects, cutting edge research, and advanced technological developments. Furthermore, the international nature of the space industry means that professionals with language skills, cultural awareness, and adaptability are highly sought after.

Navigating the process of securing international opportunities can be daunting, but with thorough research, strategic networking, and a willingness to embrace new challenges, aspiring space professionals can position themselves for success. By exploring opportunities abroad, they can gain a truly global perspective, expand their professional network, and contribute to the collective advancement of space exploration and scientific discovery.

The Future of Space Exploration

Emerging Technologies and Innovations

Hold onto your space helmets, folks, because the future of space exploration is about to get a whole lot more... well, explosive. We're talking next level thrusters, mind bending propulsion systems, and more computing power than a herd of supercomputers. Get ready for a wild ride, because the engineers of tomorrow are cooking up some seriously out of-this world ideas.

Take, for instance, the concept of nuclear thermal rockets. These bad boys use the heat generated by nuclear reactors to superheat propellant, producing thrust that could take us to the farthest reaches of the solar system in record time. Imagine zipping past Mars like a hot knife through butter, all while sipping freeze dried ice cream (the height of space cuisine, obviously).

And let's not forget about the rise of reusable launch vehicles. Forget about those pesky one time-use rockets - we're talking rockets that can land themselves, like some sort of mechanical ballerina. SpaceX has already been leading the charge on this front, but you can bet your bottom dollar that other space agencies are scrambling to catch up. Who knows, maybe in the not so-distant future, we'll have a fleet of space shuttles that can make regular trips to the Moon, Mars, and beyond.

Artificial intelligence and autonomous systems are also poised to remake space exploration. Imagine a rover that can navigate the rugged terrain of an alien world, identify interesting geological samples, and beam back stunning high definition images, all without a single human at the controls. It's the stuff of science fiction, but it's rapidly becoming science fact.

And let's not forget about the potential of biomimicry, where engineers take inspiration from nature to design inventive solutions for space travel. Who knows, maybe one day we'll have spacecraft that can "flap" their wings like birds, or life support systems that mimic the efficiency of a plant's photosynthesis. The possibilities are truly endless, and the future of space exploration is poised to be more exciting, more ambitious, and more technologically advanced than ever before.

Manned Missions to Mars and Beyond

Remember when landing on the Moon was the ultimate goal of space exploration? Well, those days are long gone, my friends. These days, the sights are set much, much higher - quite literally, in fact. The holy grail of space travel is now the long awaited journey to Mars, and the race is on to see which space agency can be the first to plant a human footprint on the red planet.

NASA, of course, has been leading the charge, with its ambitious Artemis program aiming to return astronauts to the Moon by the end of this decade, using that as a stepping stone for an eventual crewed mission to Mars. But they're not the only ones with their sights set on the Martian horizon. Private companies like SpaceX and Blue Origin are also pouring billions into developing the technology and infrastructure needed to make the dream of a manned mission to Mars a reality.

And it's not just Mars – the long term vision for space exploration goes far beyond our neighboring planet. Imagine a future where humanity has established permanent settlements on the Moon, where astronauts conduct cutting edge scientific research and extract valuable resources to fuel our continued expansion into the cosmos. Or picture a scenario where robotic probes have explored the moons of Jupiter and Saturn, uncovering the potential for life in their subsurface oceans.

Of course, undertaking such ambitious missions won't be without its challenges. The dangers of deep space radiation, the logistical hurdles of long duration spaceflight, and the sheer intricacy of establishing sustainable human presence on other worlds – these are all obstacles that will require the brightest minds in engineering, science, and technology to overcome. But if there's one thing that the history of space exploration has taught us, it's that when the human spirit is driven by curiosity and the thirst for knowledge, there's no limit to what we can achieve.

Commercialization and Privatization of Space

Hold onto your seats, folks, because the final frontier is about to get a whole lot more crowded – and a whole lot more competitive. The space industry is undergoing a seismic shift, with private companies muscling their way into a domain that was once the exclusive playground of government agencies.

Take, for instance, the rise of commercial space tourism. Companies like Blue Origin and Virgin Galactic are already offering sub orbital joy rides to deep pocketed adventure seekers, and it's only a matter of time before these experiences become more accessible to the masses. Imagine a future where a trip to the edge of space is as common as a weekend getaway – a future where the phrase "spaceflight" is no longer the stuff of science fiction, but a reality for the everyday person.

But the commercialization of space extends far beyond just tourism. Private companies are also making inroads in areas like satellite manufacturing, launch services, and even asteroid mining. Imagine a future where a network of privately owned satellites provides high speed internet access to the most remote corners of the globe, or where robotic probes extract rare earth metals from the depths of the asteroid belt, fueling the ever increasing demand for these valuable resources.

And let's not forget about the potential for private companies to play a key role in the exploration and colonization of other worlds. SpaceX's Starship program, for example, is being designed with the express purpose of establishing a permanent human presence on the Moon and Mars – something that was once the sole domain of government space agencies.

But with this shift towards commercialization and privatization comes a host of new challenges and considerations. How will the regulatory setting evolve to accommodate these new players? How can we ensure that the benefits of space exploration are distributed equitably, rather than concentrated in the hands of a few deep pocketed corporations? These are the questions that will need to be grappled with as the space industry continues to transform and evolve.

Overcoming Challenges and Setbacks

Dealing with Failures and Mistakes

In the high stakes world of space exploration, failure is never an option, right? Wrong! The truth is, even the most seasoned space professionals have faced their fair share of setbacks and missteps along the way. But here's the catch - it's not about avoiding failure altogether; it's about how you respond to it. As the old saying goes, "It's not about how many times you fall, but how quickly you get back up."

When you find yourself staring down the barrel of a failure, resist the urge to wallow in self pity or point fingers. Instead, take a deep breath, roll up your sleeves, and get to work. Analyze what went wrong, identify the lessons learned, and use that knowledge to refine your approach. Remember, mistakes are not the enemy; they're opportunities to grow and improve.

Embrace the mindset of a perpetual student. No matter how experienced you become, there's always more to learn. Approach each challenge with a curious and open mind, and don't be afraid to ask questions, seek out guidance, and collaborate with others. The path to success is paved with humility and a willingness to learn from your missteps.

Resilience and Perseverance

in the Face of Adversity

The journey to becoming a space professional is not for the faint of heart. It's a rollercoaster ride of highs and lows, triumphs and tribulations. But the true measure of a successful space engineer lies not in their ability to avoid setbacks, but in their capacity to bounce back from them.

Resilience is the superpower that will carry you through the toughest times. It's the ability to adapt, adjust, and keep moving forward, even in the face of overwhelming challenges. When the odds seem stacked against you, it's resilience that will keep you from throwing in the towel and giving up.

Develop a mindset of perseverance and determination. Remind yourself that every obstacle is a chance to prove your mettle, to stretch the limits of your capabilities, and to emerge stronger than before. Surround yourself with a support system of like minded individuals who can bolster your spirits and provide a sounding board when the going gets tough.

Remember, the greatest achievements in space exploration have often been born out of the ashes of past failures. The path to success is paved with the resilience to dust yourself off, learn from your mistakes, and keep pushing forward, no matter what the universe throws your way.

Adapting to Changes and Uncertainties

In the ever evolving terrain of space exploration, one thing is certain: change is the only constant. From the rapid advancements in technology to the shifting priorities of space agencies and organizations, the world of space engineering is a dynamic and unpredictable realm.

The ability to adapt to change and thrive in the face of uncertainty is a critical skill for any aspiring space professional. The old saying "the only constant is change" has never rung truer than in this industry, where the ground beneath your feet can shift without warning.

Embrace a mindset of flexibility and agility. Be prepared to adjust, adjust, and re strategize at a moment's notice. Grow a deep understanding of the broader context and trends shaping the space industry, so you can anticipate and respond to changes proactively.

Develop a diverse skillset and be willing to step out of your comfort zone. The space industry values professionals who can wear multiple hats and seamlessly transition between different roles and responsibilities. Stay curious, keep learning, and be ready to tackle new challenges head on.

Remember, the most successful space engineers are not those who cling to the status quo, but those who embrace the uncertainty and see it as an opportunity to shine. Adapt, evolve, and thrive in the ever changing world of space exploration, and you'll be well on your way to making your mark in this extraordinary field.

Work Life Balance and Personal Well Being

Maintaining Physical and Mental Health

As an aspiring space professional, it's easy to get caught up in the excitement and demands of your work, often at the expense of your personal well being. Remember, you're not just an engineer or scientist - you're a human being with needs that extend beyond the confines of your job. Maintaining a healthy balance between your professional and personal life is vital for long term success and fulfillment.

Start by prioritizing physical activity and a nutritious diet. It may seem like an uphill battle, with late nights and endless deadlines, but even a simple 30 minute walk or a few minutes of stretching can do wonders for your energy levels and mental clarity. Fuel your body with wholesome, brain boosting foods, and resist the temptation of the vending machine's siren call. Your mind and body will thank you for it.

Mental health is equally important, and don't be afraid to seek support when you need it. Engage in mindfulness practices, such as meditation or journaling, to help manage stress and maintain a sense of inner calm. Consider joining a local support group or speaking with a mental health professional if you're struggling with anxiety, depression, or burnout.

Remember, your well being is not a luxury - it's a necessity. When you take care of yourself, you'll be better equipped to tackle the challenges of your work and reach new heights in your space exploration career.

Achieving a Healthy Work Life Integration

In the fast paced world of aerospace engineering, it can be tempting to let your work consume every waking hour. However, it's essential to find ways to maintain a healthy balance between your professional and personal life. Establish clear boundaries and learn to say no when necessary.

Set aside dedicated time for yourself and your loved ones, and stick to it. Whether it's a weekly game night with friends, a monthly weekend getaway, or a daily evening routine to unwind, make sure to carve out this time and protect it fiercely. Resist the urge to check your work email or take that late night call – your mental and emotional well being depend on it.

Experiment with flexible work arrangements, such as remote work or flexible schedules, to create more autonomy over your time. Be open with your supervisor about your needs and work together to find a solution that works for both of you. Remember, a healthy, engaged employee is often a more productive and inventive one.

Lastly, don't forget to take regular breaks and vacations. Stepping away from the grind can help you return to your work with renewed energy and creativity. Use this time to recharge, reflect, and reconnect with the people and activities that truly matter to you.

Nurturing a Supportive Professional Network

As you navigate the demanding world of space exploration, building a strong professional network can be a game changer. Surround yourself with like minded individuals who can provide support, advice, and a sense of community.

Get involved with professional organizations, such as the American Institute of Aeronautics and Astronautics (AIAA) or the National Space Society (NSS), and attend industry events and conferences. These are excellent opportunities to connect with experienced professionals, learn from their journeys, and potentially find mentors who can guide you along the way.

Don't be afraid to reach out to your colleagues, both within your own organization and across the industry. Schedule coffee meetups or virtual check ins to discuss challenges, share successes, and provide emotional support. Offer to lend a helping hand or collaborate on projects – this reciprocity will strengthen your bonds and create a sense of camaraderie.

Remember, your professional network is not just about career advancement; it's also about cultivating meaningful relationships and a supportive community. When the going gets tough, lean on your network for inspiration, motivation, and a much needed laugh or two. Together, you'll navigate the highs and lows of the space industry, and emerge stronger for it.

Inspiring the Next Generation of Space Professionals

Mentoring and Outreach Programs

As seasoned space professionals, we hold a responsibility to pay it forward and inspire the next generation of explorers and innovators. Through mentoring and outreach initiatives, we can share our experiences, impart valuable knowledge, and ignite the passion for space exploration in young minds. One such program, the NASA Student Ambassadors, pairs undergraduate and graduate students with NASA subject matter experts, allowing them to gain firsthand perceptions into the agency's operations and the daily lives of its employees. These mentorships not only provide career guidance but also encourage a sense of belonging and community within the space industry.

Moreover, NASA's Educational Outreach programs bring the wonder of the cosmos directly to schools and communities. From visiting classrooms to hosting interactive workshops, these initiatives spark curiosity and encourage students to pursue STEM education. By showing them the tangible applications of their studies, we can help them visualize their paths to becoming future astronauts, engineers, and scientists. Imagine the thrill of a middle schooler witnessing a rocket launch or tinkering with a model rover – these experiences can leave an indelible mark and shape their academic and professional trajectories.

As mentors and outreach advocates for, we must also be mindful of encouraging diversity and inclusion. By reaching out to underrepresented communities and enabling individuals from diverse backgrounds, we can broaden the pool of talent and ensure that the space industry truly reflects the vibrant array of our society. Whether it's partnering with local organizations, participating in STEM camps for girls, or hosting virtual Q&A sessions with aspiring space professionals, our collective efforts can make a intense difference in inspiring the next generation to reach for the stars.

Encouraging STEM Education in Schools

Inspiring the next generation of space professionals starts with nurturing a strong foundation in STEM (Science, Technology, Engineering, and Mathematics) education. As industry leaders, we have a unique opportunity to collaborate with educational institutions and policymakers to improve the quality and accessibility of STEM curricula. One significant initiative is the NASA STEM Engagement program, which provides a wealth of classroom resources, lesson plans, and hands on activities that bring the excitement of space exploration into the learning environment.

By partnering with schools and districts, we can introduce new teaching methods that seamlessly integrate cutting edge technologies and real world space exploration challenges. Imagine a high school physics class designing and building model satellites, or a middle school robotics club programming autonomous rovers to navigate simulated lunar terrains. These engaging, experiential learning opportunities not only nurture critical thinking and problem solving skills but also grow a deep appreciation for the scientific principles that drive the space industry.

Moreover, we must support initiatives that address the ingrained barriers to STEM education, particularly for underserved and underrepresented communities. By supporting scholarship programs, providing mentorship opportunities, and advocating for inclusive STEM curricula, we can ensure that the path to space exploration is accessible to all. After all, the next Buzz Aldrin or Sally Ride could be a student in a resource constrained classroom, waiting to be inspired by the wonders of the cosmos and the promise of a career among the stars.

Promoting Diversity and Inclusion in the Space Industry

Diversity and inclusion are not just noble ideals; they are the keys to uncovering the full potential of the space industry. By embracing and celebrating the unique perspectives and talents of individuals from diverse backgrounds, we can encourage innovation, drive trailblazing discoveries, and inspire the next generation of space professionals.

One powerful initiative is the NASA Minority University Research and Education Project (MUREP), which partners with Historically Black Colleges and Universities (HBCUs), Hispanic Serving Institutions (HSIs), and Tribal Colleges and Universities (TCUs) to increase the participation of underrepresented groups in STEM fields. Through targeted funding, mentorship programs, and research opportunities, this project ensures that the voices and ideas of individuals from diverse communities are amplified and integrated into the space exploration community.

Beyond institutional efforts, we as industry leaders must actively promote diversity and inclusion in our own workplaces and professional networks. This might involve implementing inclusive hiring practices, sponsoring employee resource groups, or hosting workshops and panel discussions that highlight the contributions of women, minorities, and individuals with disabilities in the space industry. By dismantling biases, challenging stereotypes, and celebrating the achievements of diverse space professionals, we can create a more inclusive and welcoming environment that inspires the next generation to dream big and reach for the stars.

Copyright 2024

Silas Meadowlark

www.ingramcontent.com/pod-product-compliance
Lightning Source LLC
Chambersburg PA
CBHW030503220526
45464CB00006B/2639